AF886512

Phoenix Dactylifera Extracts

Dr. Milun Degarmo

Copyright © 2015 Dr. Milun Degarmo

All rights reserved.

Kelsiehanson Verlag

DEDICATION

To My Wonderful Readers

CONTENTS

ACKNOWLEDGMENTS

Thank you to everyone who strives to grow and help others grow

1 ABSTRACT

Phoenix dactylifera L. is a medicinal plant and it is extensively used in the prevention and cure of chronic diseases. The present study investigated the effects of antidepressant, analgesic, anti-inflammatory and anticoagulant activities in rats by *P. dactylifera* extracts. Methanolic extracts of *P. dactylifera* fruit extract (PDF), *P. dactylifera* seed extract (PDS) and *P. dactylifera* fruit and seed extract mixture (PDM) for antidepressant, analgesic, anti-inflammatory and anticoagulant activities in Sprague Dawley rats at 300, 200 and 100mg/kg body weight. Results indicated that TST and FST showed that rats struggling with efforts and immobility time were higher in PDS, while locomotory behavior and immobility time were significantly increased in PDS and PDM, respectively. All extracts showed good antidepressant activity, but PDS showed the highest value, while analgesic activity and edema reduction were demonstrated by PDF. In addition, PDM revealed significant anticoagulant activity and moderate activity in PDF. We can conclude that *Phoenix dactylifera* extracts prolong the clotting time and have the potential to reduce the platelet aggregations, depression and inflammation. Further studies needed to determine the active components responsible for pharmacological effects.

Index Terms— Analgesic activity; Anticoagulant activity; Antidepressant activity; Anti-inflammatory activity; *Phoenix Dactylifera.*

2 INTRODUCTION

Phoenix dactylifera, commonly known as Ajwa date or date palm, is a flowering plant species in the palm family, Arecaceae, which is cultivated in dry hot regions all over the world, especially in Saudi Arabia. Globally, numerous verities of dates including Sefri, Segae, Ajwa, Hilal, Khodry, Khalas, and Munifi have been originated with medicinal properties against many diseases (El Hadrami, Daayf et al. 2011). Palm of *P. dactylifera* neutralizes free radicals and inhibits the progression of many diseases. It contains various nutrients such as calcium, iron, vitamin A, vitamin B3, potassium, folic acid and other bioactive molecules including tannins, flavonoids, phenolic acids, carotenoids, triterpenoids and hydroxycinnamates (Al-Farsi and Lee 2008, El Hadrami and El Hadrami 2009, Khan, Ahmed et al. 2016). These phenolic compounds improve the immune modulators and help the human body to fight against oxidation associated chronic diseases, cardiovascular disease, inhibit the high blood pressure rate and antithrombotic effects (Al-Farsi, Alasalvar et al. 2005a, Hong, Tomas-Barberan et al. 2006). Additionally, date seeds also contain minerals such as Na, K, P, Fe, Co, Mg, Mn, Ca, Zn, Pb, Cd, Cr *etc* (Rahmani, Aly et al. 2014). It is believed that daily consumption of dates provides strength to the body, prevents from premature hair graying, reduces wrinkle formation, shiny healthy look to the skin, and also used for hardening of gums in children with dental problems (Bauza, Dal et al. 2002). Date fruit contains many phytosterols while seeds are the main reservoir for them and since the ancient era, these components have been used for the treatment of hormone-associated health problems (Mirza, Syed et al. 2019). *P. dactylifera* have several therapeutic potentials and used for the cure of many disorders such as liver, constipation, diarrhea cardiovascular disease, inflammation, analgesic and neurobiological disorders (Al-Qarawi, Abdel-Rahman et al. 2008, Al-Yahya, Raish et al. 2016). The aqueous, ethyl acetate and methanolic extract of *P. dactylifera* are known to increase the levels of enzymes cyclooxygenase 1 and 2 to prevent inflammation (Al-Yahya, Raish et al. 2016). *P. dactylifera* has traditionally been used to treat headaches and inhibit the inflammatory pathway of pain sensation by preventing prostaglandin synthesis (Sheikh, Zihad et al. 2016). The analgesic effect can be mediated by both peripherals, including prevention of mediator and/or central mechanisms that involving receptor systems. Analgesic therapy involves two key classes of analgesic drugs such as non-steroidal anti-inflammatory drugs and opioids (Bensch, Helbig et al. 2002). Both types of analgesic drugs create serious damages, such as renal dysfunction and gastrointestinal disorder (Kripalani and Kumar 2004). The diclofenac and aspirin are widely used non-steroidal anti-inflammatory drugs (Ojewole 2005). Aspirin is also an anticoagulant drug that is frequently used for the blood-thinning purpose (Buerke, Pittroff et al. 1995). Several antiplatelet medicines such as clopidogrel, prasugrel and ticagrelor are also used for this purpose. These medicines are mostly preferred when patients with coronary angioplasty and stent implantation. Anticoagulants play a vital role as mediators for the cure and inhibition of thromboembolic disorders. Unfortunately, these drugs cause several toxic life-threatening side effects like lack of reversibility, interactions with other drugs, internal bleeding, miscarriage and birth defects (Hurlen, Abdelnoor et al. 2002, Antman and Van de Werf 2004, Stone, Tonnessen et al. 2007, Bounameaux 2009). Among these anticoagulant drugs, heparin plays an essential role in the treatment, but the complication related with this anticoagulant causes internal bleeding (Hirsh and Hoak 1996), induces thrombocytopenia, low bioavailability and contamination by animal-derived pathogens as it originates from animals (Ashwini and Asha , Mourão and Pereira 1999). Eventually, there is still need for new anticoagulant agents which are effective and safe when used in cardiovascular diseases (CVDs) (Hirsh 2001). The prevalence of CVDs is continuously increases and fifteen leading diseases that are visualized in 2002-2030 by global death projection (Mathers and Loncar 2006). The main causes of mortality are stroke, hypertension, myocardial infarction, hypercholesterolemia, pulmonary embolism and deep vein thrombosis (Choi, Sapkota et al. 2013, Kang, Choi et al. 2016). *P. dactylifera* is potentially effective therapy for the cure of stuck, blocked arteries in the heart (Khan, Ahmed et al. 2016). Its extracts are widely being used to inhibit the triglycerides and cholesterol (LDL) level in the blood (Mushtaq, Kausar et al. 2017). That's why dates paste is also suggested to patients to rescue them from bypass or angioplasty treatments. Ajwa dates also contain potassium in their composition which helps in the prevention of stroke as well as comforts the nervousness. Natural products are more preferable because they have no harm on human health. *P. dactylifera* has an anticoagulant activity that serves as a noninvasive way to degrade the blood clot.

Synthetic drugs are inefficient due to their side effects and their expensive nature. Thus, there is a need for the most effective and low-cost alternative therapeutic agents from the natural source. The current investigation on the natural products for the betterment of human health and treatment of chronic diseases with no or minimal side effects is the interest of the growing world. Medicinal plants provide novel and safe therapeutic components. Therefore, the experimental application of *P. dactylifera* will explore its hidden medicinal features. This study provides scientific evidence based on the ethnomedicinal claims of the use of *P. dactylifera* for antidepressant activity, analgesic activity, anti-inflammatory activity and anticoagulant activity. For this purpose, we use three types of extracts such as *P. dactylifera* fruit extract, seed extract and their mixture to evaluate the most appropriate therapeutic agent.

1. Material and Methods

2.1. Preparation of extracts

P. dactylifera (2 kg) was purchased from the local market Madinah, Saudi Arabia. Fruits were separated from seeds and washed. Seeds were dried under shade for 24 h while fruits were left for drying in the incubator at 35-40°C for 60 days. After drying fruits and seeds were ground into a fine powder and were macerated in 99.8% methanol for 14 days with frequent shaking. After maceration, the extract was filtered with Whatman filter paper No. 01. The percentage yield of fruits and seeds was calculated which appeared to be 27.31% and 5.14% respectively. The extracts were dried under shade and were stored at -20°C till further use.

2.2. Dosage preparation

Three types of samples were prepared such as *P. dactylifera* fruit extract (PDF), *P. dactylifera* seed extract (PDS) and *P. dactylifera* fruit and seed extract mixture (PDM) 1:1. The stock solution of 100 mg/mL for extracts and 10 mg/mL for the standard drugs was prepared in 10% dimethyl sulfoxide (DMSO) with phosphate buffer saline (PBS) with pH 7.4. In the *in-vitro* assays, PBS and *in-vivo* assays normal saline in 10% DMSO served as negative control (NC) while in each assay respective standard drug served as positive control (PC). The stock of carrageenan (1%) was prepared in saline and 100 µL of this solution was administered in the left hind paw of the rat to cause inflammation that is directly associated with thrombus formation.

2.3. Animals

Sprague Dawley rats (weight 150-200 g) were grouped into positive control, negative control and treatment groups (each group *n*=7). The animals were placed under the healthy condition in aluminum crates and fed with water ad libitum and standard diet. The rats were orally administered with standard drugs and extracts as per body weight. All the extracts were orally administrated with a concentration of 300, 200 and 100 mg/kg, and 10, 5 and 2.5 mg/kg for standard drug according to rat body weight. In each experiment, rats were divided into controls (NC and PC) and extract groups (PDF, PDF and PDM).

2.4. Tail suspension test

Tail suspension test (TST) based on stress induction was used to investigate the antistress effects of the *P. dactylifera* extracts by the previously reported method (Poleszak, Szopa et al. 2016). One day before the experiment, rats were hanged for 5 min by tails with a thread having the support of iron stand to swung rat body in the air. Initially, rats struggled to climb to the support by moving upward but after some time rats stopped struggling that state was considered as "given up" which represents the depression state. The immobility time was recorded with the help of the camera on the top of tail suspension. After 24 h of first session rats were orally administered with extracts and standard drug (Fluoxetine-HCl) accordingly. Then the test was performed to determine the antidepressant effect of plant extracts. Kinoscope_3.0.4 supported by Preclinical Data Forum was used to assign a score to all events happened during the test (Kokras, Baltas et al. 2017). The experiment was performed in replicates and percentage activity was calculated by the following formula.

$$\text{Percentage antidepressant activity} = [((M - N)/M) \text{x } 100]$$

Where 'M' is average immobility of negative control while 'N' is immobility value of each rat of all other four groups.

2.5. Forced swimming test

The forced swimming test (FST) was used to determine the antidepressant activity of *P. dactylifera* extracts by following the same process as described previously with minor modifications (Slattery and Cryan 2012). The test was performed in a glass cylinder of 60 cm height and 20 cm × 20 cm square area; half-filled with water. In the pre-test, session rats were exposed to swimming-stress for 10 min one day before the test. Next day all groups were feed with respective dose and standard drug (Fluoxetine-HCl) 1 h before test performance and the camera was placed to record the event. Rat behavior was judged by its score that assigned by observers according to the state of immobility and lime movement to kept head on outside of water. Average of three scores was considered as final score (Gu, Zhou et al. 2014). Kinoscope_3.0.4 supported by Preclinical Data Forum was used to count different events happened during the test. This software provided the duration of each event, assigned a score to the event, the number of times event took place and frequency of event (Kokras, Baltas et al. 2017). Finally, the percentage of immobility was calculated by the following formula.

$$\text{Percentage activity} = \left[\left(\frac{NC-NS}{NC}\right) \text{x } 100\right]$$

Where "NC" is the average immobility of negative control and "NS" is the average immobility of the experimental group.

2.6. Open field test

P. dactylifera extracts were evaluated by open field test (OFT) to determine the locomotor activity level and willingness to explore in animals by previous method (Stewart 2012). Open field assay apparatus has consisted of hardboard covered with white sheet ground and four walls. The square floor of 75 × 75 cm was divided by black lines in 16 equal portions and walls were of 30 cm height. Test samples and standard drug (Fluoxetine-HCl) were administered 1 h before the test performance and each rat was placed one-by-one in the center of the square board for 5 min and its behavior was recorded with the help of the camera on the top. Evaluation of antidepressant behavior number, the pattern of square crossed, total locomotion, and frozen events were determined. Evaluation of final score was based on the comparative study of behavioral events of control and treated groups and the percentage activity was calculated by the following formula.

$$\text{Percentage activity}= \left[\left(\frac{\text{A}-\text{B}}{\text{A}}\right) \text{x } 100\right]$$

Where 'A' is average immobility of negative control while 'B' is the average immobility of the test group.

2.7. Hole Board Test

Hole board test (HBT) was used to examine stress and neophilic behavior in rats by the previously reported method (Can, Özkay et al. 2010). HBT apparatus was consisted of a hard board of 75 × 75 cm, containing 16 equal size holes evenly marked inboard and the ground of hardboard was bounded by four 30 cm walls. Individually rats were placed in the center of the board and were allowed to freely explore the holes on board. A camera was adjusted by focusing on board to record all the events. Parameters were set for evaluation included; the number of explored holes, head-dips, and latency period before head-dips for the first time. All rats were explored to the apparatus 1 day before the test for training purposes. One hour before the experiment, rats were administered orally according to their grouping, and all events were recorded by camera for 5 min. ToxTrac v_2.83 supported by Experimental Lakes Area Inc was used for evaluation of each video, it provided the information of mobility rate, frozen events, immobility time, latency period, number of head dips and explored area. Fluoxetine-HCl was used as a positive control while 10% DMSO was used as the negative control. The experiment was performed in replicates and percentage activity was calculated by the following formula.

$$\text{Percentage antidepressant activity}= \left[\left(\frac{\text{X}-\text{Y}}{\text{X}}\right) \text{x } 100\right]$$

Where 'X' is average immobility of negative control while 'Y' is the immobility of test groups.

2.8. Writhing assay

To evaluate the peripheral analgesic effect of extracts, the writhing syndrome model was used as described previously (Mondal, Saha et al. 2014). The rats were administrated by PDF, PDS, PDM, PC (Morphine), and NC (DMSO) to their respective groups. Thirty minutes after the dosage, rats were intraperitoneally injected with acetic acid to induce writhing and were placed on a flat surface to determine the number of writhes. Data was recorded by counting writhing movements which is categorized by abdominal contraction and stretching of hind limbs. The percentage activity was measured by the following formula:

$$\text{Percentage inhibition} = \left[\left(\frac{\text{Nc} - \text{Ns}}{\text{Nc}}\right) \text{x } 100\right]$$

Where" Nc" is the mean value of the saline group and Ns is the mean value of the extract group.

2.9. Hot plate assay

A hot plate assay was used to evaluate the analgesic effect of extracts by the previously described method (Ismail, Dilshad et al. 2016). Morphine was used as a positive control (PC). Before treatment, initial latency time was determined by measuring animal response such as licking paws and jumping off by placing them on the hot plate at 55 ºC with the cutoff period of 20 sec to avoid the injury to rat paw. The latency was recorded after 30 min of oral administration of various extracts at the time interval of 1, 2 and 3 h of each rat by placing on a hot plate under the same condition, and data was recorded as endpoint by observing animal jumping or licking response of hind paw. Percentage analgesic activity was measured by the formula:

$$\text{Percentage analgesic activity} = \left[\left(\frac{\text{Tf} - \text{Ti}}{\text{Ti}}\right) \text{x } 100\right]$$

Where "Ti" is an initial latency time and "Tf" is endpoint latency time.

2.10. Paw edema assay

The anti-inflammatory activity of *P. dactylifera* extracts using carrageenan-induced hind paw edema assay as reported (Ismail, Dilshad et al. 2017). Diclofenac potassium was used as a positive control. After 1 h of oral dosage, edema was induced by 100 µL of carrageenan (1% in saline) injection into the sub plantar region of the left hind paw. The paw volume was calculated instantly before and after the carrageenan injection by Plethysmometer (UGO Basile 7140). After that, the paw volume was measured at 0, 1, 2 and 3 h and the percentage edema inhibition was measured by the formula:

$$\text{Percentage inhibition} = \left[\left(\frac{\text{A} - \text{B}}{\text{A}}\right) \text{x } 100\right]$$

Where "A" is edema of control rats and "B" is edema of treated rats.

2.11. Histamine assay

Histamine assay was performed to investigate the anti-inflammatory effect of *Phoenix dactylifera* extracts by reported method (Sajid, Khan et al. 2017). In this assay, edema was induced by injecting 0.1 mL of histamine (1 mg/mL) into the left hind paw of each rat. Chlorpheniramine maleate was used as a positive control. Each extract was given oral dose, edema was measured and calculated at 0, 1, 2 and 3 h respectively by Plethysmometer (UGO Basile 7140). The values were taken in replicates and the percentage edema inhibition was measured by the following formula:

$$\text{Percentage inhibition} = \left[\left(\frac{A - B}{A}\right) x\ 100\right]$$

Where "A" is edema of control rats and "B" is edema of treated rats.

2.12. Sheep blood assay

Sheep blood was collected from the slaughterhouse and platelet-poor plasma (PPP) was prepared as described previously (Dutta, Gogoi et al. 2015) which assesses hemolysis and protein denaturation in the erythrocyte. 100 μL of PPP and 2 μL of the sample (PDF, PDS, and PDM) were added in 96-well plates followed by incubation for 3 min at 37°C. Then 40 μL of 250 mM CaCl2 was added in the reaction mixture to initiate the clot formation (Saikia, Thakur et al. 2011). The negative control was prepared with 1x PBS (pH 7.4) while the positive control group was prepared by aspirin instead of the test sample. Right after the addition of CaCl2 absorbance was measured at 540 nm using a microplate reader (Diamate biotechnologies Ltd, Dia 710) with time interval ranges from 1 to 5 min. The experiment was further extended by taking absorbance after every 30 min to 3 h and percentage activity was calculated by using the following formula.

$$\text{Percentage activity} = \left[\left(\frac{A\text{-}B}{A}\right) x\ 100\right]$$

In this formula, "A" represents the average absorbance activity of extract while "B" represents the average absorbance activity of the PBS control group.

2.13. Human antiplatelet assay

The antiplatelet assay was performed by the previously reported method (Dhurat and Sukesh 2014) to investigate the anticoagulant behavior of extracts. In this assay, platelet-rich plasma (PRP) was prepared from human blood. The 2 μL of each *P. dactylifera* extracts (PDF, PDS, and PDM), Aspirin (positive control), and PBS (negative control) were added separately in 96-well plate followed by addition of 100 μL of PRP in each well. The plate was incubated for 3 min at 37°C and absorbance was measured at 540 nm with the help of a microplate reader (Diamate Biotechnologies Ltd, Dia 710). Absorbance was consecutively measured after 1 to 5 min and then measured after every 30 min to 3 h. The experiment was performed in triplicate and percentage activity was calculated according to the following formula.

$$\text{Percentage activity} = \left[\left(\frac{NC\text{-}NS}{NC}\right) x\ 100\right]$$

In this formula, "NC" represents the average absorbance of PBS control group and "NS" represents the average absorbance values of the sample groups.

2.14. Rat capillary assay

Rat capillary method was used according to the previously reported method (Ismail and Mirza 2015) to determine the time delay in clot formation after the treatment of extracts. After 1 h of dosage, the rat tail was pierced with the help of lancet and the tail blood was filled in capillary tubes and placed in a water bath at 37ºC. After 60 s the small parts of the capillary tube were fragmented each 10 s until the thread of fibrin is visualized between the pierced ends of the tube. Aspirin was used as positive control and saline in 10% DMSO served as the negative control. Clot formation time was recorded in seconds and in the form of International Normalized Ratios (INR) (Lee, Lee et al. 2016). The INR values were calculated by the following formula.

$$INR=\left(\frac{PT\ sample}{PT\ control}\right)^{ISI}$$

In this equation, PT represents the prothrombin time (clotting time) and ISI represents the International sensitivity index as per WHO guideline.

2.15. Carrageenan induced thrombosis assay

The anticoagulant activity of *P. dactylifera* was determined by carrageenan-induced thrombosis in rat assay as reported previously (Yan, Yan et al. 2009). After 1 h of dosage, thrombus formation was initiated into the sub plantar region of left hind paw by injecting 100 µL of carrageenan (1% in saline). The rat paw was measured before and after the carrageenan injection to determine the thrombus size by using Vernier caliper. Aspirin and 10% DMSO served as positive and negative controls, respectively. Readings for thrombin size were taken with regular interval of 1 h up to 3 h and the percentage inhibition for thrombus size was determined by the following formula:

$$\text{Percentage inhibition} = \left[\left(\frac{X-Y}{X}\right) \times 100\right]$$

In this formula, "X" represents the thrombus size of control rats and "Y" represents the thrombus size of treated rats.

2.16. Statistical Analysis

The normality of raw data was analyzed by Minitab-17 and then the outlier test was performed to detect any outlier in the sample. After that data was analyzed by Analysis of Variance (ANOVA) followed by Tukey's test using GraphPad Prism 5.0 software. Results are represented as mean ± SD and $p < 0.05$ is considered to be significant.

3. Results and Discussion

3.1. Antidepressant activity

The antidepressant activity of *P. dactylifera* was investigated by tail suspension test using animal jumping and grooming time. In addition, animal given-up time was recorded after several movements, while immobility time was considered to be zero activity. The highest struggling and hanging activity were showed by PDS 89.7±3.3 sec, while immobility duration was 26.9±2.1 sec (Table 1). PDS and PDM have closest values to each other which determined the neutral effect of fruit and seed mixture. The percentage activity of *P. Dactylifera* extracts showed 71% antidepressant activity by PDS, while PC 74% as compared to NC (Fig. 1A). Results showed that PDS has the highest antidepressant activity as compared with negative control, which can be explained due to the presence of antioxidants (Zhang, Aldosari et al. 2013). Previous reports showed the reduction of oxidative stress causes depression and antioxidant properties of flavonoids (Pietta 2000). In addition, anthocyanin plays an important role in the repair of nerve damages by providing potential antioxidant activity. *P. dactylifera* contain a small amount of anthocyanin which can perform the antidepressant activity by repairing the antioxidant damages (Al-Farsi, Alasalvar et al. 2005a).

Moreover, FST is an efficient method to check the effectiveness of antidepressant drugs without effecting the locomotory activity (Detke, Rickels et al. 1995). FST was performed to evaluate the antidepressant activity of *P. dactylifera*. Our results showed that PDS exhibited significant climbing time 88.2±1.76 sec with lowest immobility time 18.61±1.8 sec while PDM demonstrated moderate climbing 78.39±1.77 sec and immobility time 20.07±4.49 sec (Table 1). In addition, percentage activity was PDS (70%), PDF (48%) and PDM (66%) activity as compared with negative control while PC displayed 75% activity (Fig. 1B). Previous studies have been reported that several glycosides containing plants can provide potential antidepressant activity (Khan, Amin et al. 2018). Numerous glycosides target the variation of a brain-derived neurotrophic factor in the hippocampus that links the depression disorder. Glycosides link to the flavonoid, anthraquinone, steviol, phenolic, chromone, saponins and iridoid (Aydemir, Yalcin et al. 2006). Flavonoids are responsible for the reduction of oxidative stress and resulted in antidepressant activity (Tapas, Sakarkar et al. 2008). The presence of such compounds suggests the possible mechanism for the antidepressant activity of *P. dactylifera*. Our results indicated that seeds extract to have significant antidepressant activity by reducing the immobility duration as compared to the negative control.

Next, OPT was performed to investigate the antidepressant activity of *P. dactylifera* in terms of general locomotor activity levels and willingness to explore. We found that the highest antidepressant activity percentage was showed by PDS (60%) as compared to the negative control (Fig. 1C). Whereas, the highest mobility rate was showed by PDS 81.76±1.4% with mean explored areas 13.29±1.1. In contrast, the immobility duration for PDF 73.5±4.5 sec, PDS 54.71±4.3 sec, and PDM 87.71±3.4 sec (Table 1). In addition, PDF showed the lowest number of frozen events 11.67±1.8, whereas PDM showed higher immobility which represents the antagonistic correlation of mixture (Table 1). OFT has been reported as a suitable tool for the determination of ethological behavior analysis which allows the investigation of both specific and non-specific effects of anxiolytic drugs (Crawley 1985).

Further, the antidepressant activity of *P. dactylifera* in terms of the number of head-dips and exploration time was investigated by HBT. The results indicated that all extracts revealed significant antidepressant effects in terms of the number of head dips, latency time, explored area and immobility time as compared with control. The highest mobility rate was displayed by PDM 50.57±4.7%, while numbers of frozen events were 6.71±1.5 in PDM extract (Table 1). Next, we found the lowest latency time by PDM 4.71±1.5 s, while the highest number of head dips was exhibited by PDS 9.29±2.6, as compared with NC (Table 1). Overall in HBT, antidepressant activity was showed by PDS 70%, PDF 65%, and PDM 60% (Fig. 1D). Previously, behavior studies reported that running, walking, exploration, and rearing can happen in response to the occurrence of any stressful inducement (Armario and Garcia-Marquez 1987). In the present study, PDS has shown the highest number of head-dips, mobility rate, explored the area, as well as the lowest latency period, increases the antidepressant effect. *P. dactylifera* contains antioxidants like flavonoids which are responsible for the reduction of oxidative stress (Zhang, Aldosari et al. 2013). Environmental and emotional factors indirectly cause oxidative stress which results in nerve damages in the brain and expressed as depression disorder. *P. dactylifera* has the unique potential to cure such damages due to its phytochemical composition (Khalid, Khalid et al. 2017, Saryono, Proverawati et al. 2017). This suggested that *P. dactylifera* seeds have antidepressant potential.

3.2. Analgesic activity

The writhing assay was used to evaluate the effect analgesic activity of *P. dactylifera* by acetic acid-induced writhing assay. In our results, significant writhes number 13.71±1.71 and analgesic activity (80%) was showed in PDF extract (Table 2 and Fig. 2A). *P. dactylifera* is reported to contain phenolics such as cinnamic acid, caffeic acid, and ferulic acid etc., carotenoids, sterols, flavonoids, and their glycosides which decrease the number of writhes (Mansouri, Embarek et al. 2005). Studies have reported that flavonoids and polyphenols components in Safawy and Sukkary date extract were involved in decreasing the number of writhes (Murata, Ushikubi et al. 1997). Furthermore, plant extracts are known to inhibit the prostaglandin formation through which they block the inflammatory pathway of pain sensation (Sheikh, Zihad et al. 2016). Reports have also suggested that the higher content of *trans*-ferulic acid in Ajwa dates as compared to other dates verities have shown stronger analgesic activity in acetic acid-induced writhing test in rats (Lv, Zhang et al. 2013). Ajwa dates are reported to contain phenolics such as cinnamic acid, caffeic acid, and ferulic acid, etc., carotenoids, sterols, flavonoids and their glycosides (Mansouri, Embarek et al. 2005). The presence of these compounds suggests the possible mechanism for the nociceptive activity of Ajwa dates.

Next, the analgesic activity of *P. dactylifera* extracts and latency time was determined as a time-dependent manner in hot plate assay (Table 2). We found that 80% analgesic activity was showed by PDF extract as compared to NC (Fig. 2B). Previously, researchers have reported that *P. dactylifera* contains alkaloids, sterols, amino acids, and flavonoids that inhibit pain (Zulfiker, Rahman et al. 2010). The hot plate method is one of the easiest and suitable methods for the determination of analgesic activity in rats. The hot plate method represents the centrally acting mechanism of pain which usually raises the pain threshold of rats to pressure or heat (Khan, Saeed et al. 2011). Researchers have found that sterols and flavonoid components in different varieties of date Boufgous, Bousrdon and

Majhoul were involved in pain inhibition and decreasing analgesia (Jain and Priyadarshan 2009). From the previous study, it was determined that Ajwa date extracts also contain such flavonoid which target prostaglandins and are involved in late phase of pain discernment and analgesic activity. The alkaloids in Ajwa dates are also reported for inhibition of pain perception (Abdullahi, Ilyas et al. 2013). It was reported that Ajwa dates fruits and seed extracts are enriched with sterols and amino acid components and these components are responsible for necrotic analgesia (Awaad, El-meligy et al. 2011). In one study, the ethanolic extract of Ajwa dates was evaluated for its analgesic potential and researchers have found that these extracts exhibit central acting analgesic activity and are also involved in the prolongation of hot plate latency time (Zulfiker, Rahman et al. 2010). Previous reports also suggest that Ajwa dates contain steroids and tannins which have a contribution to analgesic activity (Malairajan, Geetha et al. 2006). Overall, results suggest that Ajwa dates are enriched with phenylpropanoids which has maximum analgesic activity in thermal hyperalgesia test in rats and its polar soluble components are involved in the inhibition of both central and peripheral pathway of pain.

3.3. Anti-inflammatory activity

Carrageenan induced paw edema assay was used to measure the anti-inflammatory potential of *P. dactylifera* extracts. The results of edema volume at different time intervals are presented in Table 3. The maximum percentage inhibition of edema was exhibited by PDF (85%) extract as compared with NC (Fig. 3A). Our results showed that *P. dactylifera* extract inhibits the release of histamine and PG synthesis. Studies demonstrate that *P. dactylifera* has gallic acid, phenolic acid, methanolic, and ethyl acetate extracts which are known for potent anti-inflammatory activities (El Abed, Chakroun et al. 2018). Previous studies demonstrate that Ajwa dates contain caffeic acid, p-coumaric acid, syringic acid, ferulic acid, gallic acid and rutin in various concentrations (Eid, Al-Awadi et al. 2013). Gallic acid is a major phenolic acid present in Ajwa date is known for potent anti-inflammatory activity. It inhibits the releases of histamine pro-inflammatory cytokines from activated mast cells of human and has membrane stabilizing potential (Di Rosa, Giroud et al. 1971). During the evaluation of Ajwa dates methanolic and ethyl acetate extracts, scientists have found that these extracts inhibit the lipid peroxidation of cyclooxygenase enzyme (COX-1 and COX-2), transcription factor (LOX and NF-κB) and also have a role in the reduction of paw edema and plasma fibrinogen (Ahmed, Bano et al. 2015). Presence of catechin and epicatechin glycosides in Ajwa dates extracts have been reported that they are involved in the anti-inflammatory activity in rats by inhibiting the nitric oxide (NO) synthesis whose production was promoted by carrageenan injection and prevent the release of LPS-induced prostaglandin E2 (García, Salazar et al. 2012).

In addition, extracts of *P. dactylifera* were evaluated using histamine assay to determine the anti-inflammatory effect after carrageenan injection (Di Rosa, Giroud et al. 1971). The change in edema volume at different time-intervals was determined in all groups (Table 3). We found the most significant activity of PDM (80%) in the rat model as compared with NC (Fig. 3B). It was demonstrated that *P. dactylifera* extracts have phthalates and fatty acids such as linoleic, stearic, and oleic acids which inhibit COX-1 and COX-2 enzymes of the inflammatory pathway (Ramsewak, Nair et al. 2001). Previously, the lyophilized extract of Ajwa dates was evaluated which showed the reduced expression of pro-inflammatory cytokines such as IL-10, TNF α, IL-6, and apoptotic markers such as Bax and caspase-3 (John, Muller et al. 2007). The lyophilized water extract of Ajwa dates yielded proteinaceous fraction K which has antioxidant and anti-inflammatory activity (Al-Yahya, Raish et al. 2016). It was demonstrated that Ajwa date extracts have phthalates and fatty acids such as linoleic, stearic, and oleic acids which inhibit COX-1 and COX-2 enzymes of the inflammatory pathway (Ramsewak, Nair et al. 2001). Studies reported that Ajwa dates have multiple modulator molecules which can efficiently help in the reduction of pain like glycosides inhibit cyclooxygenase enzyme of the inflammatory pathway to reduce edema (Bennett, Stamford et al. 1988), quercetin, luteolin and triterpenoids flavonoids inhibit both cyclo and lipoxygenase pathway of inflammation (Safayhi and Sailer 1997). Our results revealed that the administration of Ajwa dates inhibited the edema during all phases of inflammation, which is probably inhibition of different aspects and chemical mediators of inflammation.

3.4. Anticoagulant activity

The anticoagulant activity of *P. dactylifera* was evaluated by sheep blood assay. Our results indicated that the behavior of PDM and PDF was almost similar to PC (Fig. S1A). Percentage clot formation of sheep blood assay showed PDF 62.24% and PDM 62.15% inhibition (Fig. 4A). It is stated that all the anticoagulant fractions contained an abundant amount of carbohydrates content and polyphenolic components (Khoo, Abas et al. 2014). The *P. dactylifera* extract also contains 80% sugar content in its composition and negatively charged polyphenols are also known to exhibit the anticoagulant activity. Previously, it has been revealed that platelet functional role is very vital for the thrombotic risk. The decreased platelet aggregation due to the platelet-poor plasma by the *P. dactylifera* extracts can also be recognized by the significant enhancement in the platelet cytosolic level that is cyclic adenosine monophosphate (cAMP) pathway (Halbrügge, Eigenthaler et al. 1992). The encouragement of platelet is inhibited by the intracellular associated cAMP and cyclic guanosine monophosphate (cGMP) pathways (Walter, Eigenthaler et al. 1993).

Subsequently, the anticoagulant activity of *P. dactylifera* was determined by antiplatelet assay. The results of *P. dactylifera* induce a change in the absorbance with a time-dependent manner (Fig. S1B). Further, clotting proportion and coagulation activity comparison by PRP showed that PDM extract showed highly significant anticoagulant activity. In addition, anticoagulant activity percentage was higher in PDM (82%) as compared to control (Fig. 4B). Previously antiplatelet activity of *C. paradise* plant extract estimated that flavonoids may change the coagulation and fibrinolysis processes (Mallick, Khan et al. 2016). Similarly, another study revealed that *Plinia cauliflora* (Mart.) leaves extract contain tannins and flavonoids (flavones, flavanones, isoflavones, and chalcones) in its composition (Tanelli[1], Belardo et al. 2017). These components are responsible for the alteration in the hemostasis pathways. Therefore, we suggest that these special flavonoids are responsible for the anticoagulant activity in *P. dactylifera* extracts.

Next, we performed rat capillary assay to evaluate the anticoagulant ability of *P. dactylifera*. The most significant activity was showed by PDF extract with coagulation time 299.14±4.5 sec (Fig. 1SC). Additionally, the international normalized ratio (INR) was used to monitor the balance between the risks of excessive bleeding against the risk of clotting. The INR results exhibited that PDF extract has a significant anticoagulant effect (Fig. 4C). The previous study on the Khalas dates has proposed an anticoagulant effect due to their polyphenolic compositions (Hadi, Elderbi et al. 2010). In another study, it was testified that antioxidants can stabilize the hematological and blood coagulation turbulences, oxidative stress, as well as hepatorenal damages (Ashour 2014). Thus, *P. dactylifera* contains a high amount of antioxidants that can be served strong anticoagulant agent with antioxidant constituents (Vayalil 2002).

Further, the anticoagulant activity of *P. dactylifera* was investigated by carrageenan-induced thrombosis assay in the rat. The results of percentage inhibition in a time-dependent manner of thrombus size after carrageenan injection found that PDF showed 86% inhibition in thrombus size as compared to control (Fig. 4D). While the measurement of thrombus size values is given in (Table 4). The pharmacological study of carrageenan is responsible for inflammation in the local blood vessels which involve in thrombus formation by interleukin-1 as well as tumor necrosis factor (Patil, Mahajan et al. 2019). These factors are the reason for the breakdown of healthy endothelial cells that regulate the fibrinolysis as well as hemagglutination mechanisms (Wang, Li et al. 2000). Researchers have explained that the decrease in thrombus size after carrageenan injection in *E. neriifolia* plant extracts is due to the presence of components like polyphenols and flavonoids (Hasan, Ganeshpurkar et al. 2014).

4. Results

In conclusion, our results indicate that *P. dactylifera* extracts (PDS, PDF and PDM) have significant antidepressant, analgesic, anti-inflammatory and anticoagulant activities. The prominent antidepressant activity was found in the extract of *P. dactylifera* seed. The highest analgesic activity was found in the PDF as compared to PDS which provide both peripheral and central analgesic effects. The results indicated that there was a maximum inhibition of edema in PDF as compared to PDS and PDM in all phases of inflammation. Additionally, results of *in-vivo* assays showed that PDM extract showed a synergetic effect in antiplatelet, while the highest *in-vitro* anticoagulant effect was revealed by PDF. Overall our results represent that *P. dactylifera* extracts showed highly significant anticoagulant activity *in-vitro* and *in-vivo* models. The research further suggests identifying and isolating that specific component that is responsible for all these acts of *P. dactylifera*.

References

Abdullahi, M. N., N. Ilyas and H. Ibrahim (2013). "valuation of Phytochemical Screening and Analgesic Activity of Aqueous Extract of the Leaves of Microtrichia perotitii Dc (Asteraceae) in Mice Using Hotplate Method. ." Medicinal Plant Research **3**: 37-43.

Ahmed, A., N. Bano and M. Tayyab (2015). "Phytochemical and Therapeutic Evaluation of Date (Phoenix dactylifera). A Review." **9**.

Al-Farsi, M., C. Alasalvar, A. Morris, M. Baron and F. Shahidi (2005a). "Compositional and sensory characteristics of three native sun-dried date (*Phoenix dactylifera L.*) varieties grown in Oman." *Journal of Agricultural and Food Chemistry* **53**(19): 7586-7591.

Al-Farsi, M. A. and C. Y. Lee (2008). "Nutritional and functional properties of dates: a review." Critical reviews in food science and nutrition **48**(10): 877-887.

Al-Qarawi, A., H. Abdel-Rahman, H. Mousa, B. Ali and S. El-Mougy (2008). "Nephroprotective action of Phoenix dactylifera. in gentamicin-induced nephrotoxicity." Pharmaceutical Biology **46**(4): 227-230.

Al-Yahya, M., M. Raish, M. S. AlSaid, A. Ahmad, R. A. Mothana, M. Al-Sohaibani, M. S. Al-Dosari, M. K. Parvez and S. Rafatullah (2016). "'Ajwa' dates (Phoenix dactylifera L.) extract ameliorates isoproterenol-induced cardiomyopathy through downregulation of oxidative, inflammatory and apoptotic molecules in rodent model." Phytomedicine **23**(11): 1240-1248.

Al-Yahya, M., M. Raish, M. S. AlSaid, A. Ahmad, R. A. Mothana, M. Al-Sohaibani, M. S. Al-Dosari, M. K. Parvez and S. Rafatullah (2016). "'Ajwa'dates (Phoenix dactylifera L.) extract ameliorates isoproterenol-induced cardiomyopathy through downregulation of oxidative, inflammatory and apoptotic molecules in rodent model." Phytomedicine **23**(11): 1240-1248.

Antman, E. M. and F. Van de Werf (2004). "Pharmacoinvasive therapy: the future of treatment for ST-elevation myocardial infarction." *Circulation* **109**(21): 2480-2486.

Armario, A. and C. Garcia-Marquez (1987). "Tricyclic antidepressants activate the pituitary-adrenal axis in the rat. Tolerance to repeated drug administration." *European Journal of Pharmacology* **140**(2): 239-244.

Ashour, T. H. (2014). "Preventative effects of caffeic acid phenyl ester on cadmium intoxication induced hematological and blood coagulation disturbances and hepatorenal damage in rats." ISRN hematology **2014**.

Ashwini, U. and S. Asha "A Preliminary Novel Findings on Invitro Anti-Coagulant Activity of Acalypha Indica Linn Leaf Extracts."

Awaad, A. S., R. M. El-meligy, S. A. Qenawy, A. H. Atta and G. A. Soliman (2011). "Anti-inflammatory, antinociceptive and antipyretic effects of some desert plants." Journal of Saudi Chemical Society **15**(4): 367-373.

Aydemir, C., E. S. Yalcin, S. Aksaray, C. Kisa, S. G. Yildirim, T. Uzbay and E. Goka (2006). "Brain-derived neurotrophic factor (BDNF) changes in the serum of depressed women." *Progress in Neuro-Psychopharmacology and Biological Psychiatry* **30**(7): 1256-1260.

Bauza, E., C. F. Dal, A. Berghi, G. Oberto, D. Peyronel and N. Domloge (2002). "Date palm kernel extract exhibits antiaging properties and significantly reduces skin wrinkles." *International Journal of Tissue Reactions* **24**(4): 131-136.

Bennett, A., I. F. Stamford, I. A. Tavares, S. Jacobs, F. Capasso, N. Mascolo, G. Autore, V. Romano and G. Di Carlo (1988). "The biological activity of eugenol, a major constituent of nutmeg (Myristica fragrans): Studies on prostaglandins, the intestine and other tissues." Phytotherapy Research **2**(3): 124-130.

Bensch, S., A. J. Helbig, M. Salomon and I. Seibold (2002). "Amplified fragment length polymorphism analysis identifies hybrids between two subspecies of warblers." Molecular Ecology **11**(3): 473-481.

Bounameaux, H. (2009). "The novel anticoagulants: entering a new era." *Swiss Medical Weekly* **139**(5-6): 60-64.

Buerke, M., W. Pittroff, J. Meyer and H. Darius (1995). "Aspirin therapy: optimized platelet inhibition with different loading and maintenance doses." *American Heart Journal* **130**(3): 465-472.

Can, Ö. D., Ü. D. Özkay, N. Öztürk and Y. Öztürk (2010). "Effects of hawthorn seed and pulp extracts on the central nervous system." *Pharmaceutical Biology* **48**(8): 924-931.

Choi, J.-H., K. Sapkota, S.-E. Park, S. Kim and S.-J. Kim (2013). "Thrombolytic, anticoagulant and antiplatelet activities of codiase, a bi-functional fibrinolytic enzyme from Codium fragile." *Biochimie* **95**(6): 1266-1277.

Crawley, J. N. (1985). "Exploratory behavior models of anxiety in mice." *Neuroscience & Biobehavioral Reviews* **9**(1): 37-44.

Detke, M. J., M. Rickels and I. Lucki (1995). "Active behaviors in the rat forced swimming test differentially produced by serotonergic and noradrenergic antidepressants." *Psychopharmacology* **121**(1): 66-72.

Dhurat, R. and M. Sukesh (2014). "Principles and Methods of Preparation of Platelet-Rich Plasma: A Review and Author's Perspective." J Cutan Aesthet Surg **7**(4): 189-197.

Di Rosa, M., J. Giroud and D. Willoughby (1971). "Studies of the mediators of the acute inflammatory response induced in rats in different sites by carrageenan and turpentine." The Journal of pathology **104**(1): 15-29.

Di Rosa, M., J. P. Giroud and D. A. Willoughby (1971). "Studies on the mediators of the acute inflammatory response induced in rats in different sites by carrageenan and turpentine." J Pathol **104**(1): 15-29.

Dutta, S., D. Gogoi and A. K. Mukherjee (2015). "Anticoagulant mechanism and platelet deaggregation property of a non-cytotoxic, acidic phospholipase A2 purified from *Indian cobra* (Naja

naja) venom: inhibition of anticoagulant activity by low molecular weight heparin." *Biochimie* **110**: 93-106.

Eid, N. M., B. Al-Awadi, D. Vauzour, M. J. Oruna-Concha and J. P. Spencer (2013). "Effect of cultivar type and ripening on the polyphenol content of date palm fruit." J Agric Food Chem **61**(10): 2453-2460.

El Abed, H., M. Chakroun, Z. Abdelkafi-Koubaa, N. Drira, N. Marrakchi, H. Mejdoub and B. Khemakhem (2018). "Antioxidant, Anti-Inflammatory, and Antitumoral Effects of Aqueous Ethanolic Extract from Phoenix dactylifera L. Parthenocarpic Dates." Biomed Res Int **2018**: 1542602.

El Hadrami, A., F. Daayf and I. El Hadrami (2011). Date palm genetics and breeding. Date Palm Biotechnology, Springer**:** 479-512.

El Hadrami, I. and A. El Hadrami (2009). Breeding date palm. Breeding plantation tree crops: tropical species, Springer**:** 191-216.

García, M., F. Salazar, C. Ibarra, F. Michelangeli, M. Ruiz, T. Peter and J. Villamizar (2012). "Anti-inflammatory effects of Byrsonima crassifolia." Planta Medica **78**: PI130.

Gu, X., Y. Zhou, X. Wu, F. Wang, C.-Y. Zhang, C. Du, L. Shen, X. Chen, J. Shi and C. Liu (2014). "Antidepressant-like effects of auraptenol in mice." Scientific reports **4**: 4433.

Hadi, A., M. A. Elderbi and A. Mohamed (2010). "Effect of gum arabic on coagulation system of albino rats." Int J PharmTech Res **2**: 1762-1766.

Halbrügge, M., M. Eigenthaler, C. Polke and U. Walter (1992). "Protein phosphorylation regulated by cyclic nucleotide-dependent protein kinases in cell extracts and in intact human lymphocytes." *Cellular Signalling* **4**(2): 189-199.

Hasan, M., A. Ganeshpurkar, D. Bansal and N. Dubey (2014). "Protective effect of Euphorbia neriifolia extract on experimentally induced thrombosis in murine model." *Nigerian Journal of Experimental and Clinical Biosciences* **2**(2): 86-88.

Hirsh, J. (2001). "New anticoagulants." American heart journal **142**(2): S3-S8.

Hirsh, J. and J. Hoak (1996). "Management of deep vein thrombosis and pulmonary embolism: a statement for healthcare professionals from the council on thrombosis (in consultation with the council on cardiovascular radiology), American Heart Association." *Circulation* **93**(12): 2212-2245.

Hong, Y. J., F. Tomas-Barberan, A. A. Kader and A. E. Mitchell (2006). "The flavonoid glycosides and procyanidin composition of Deglet Noor dates (*Phoenix dactylifera*)." *Journal of Agricultural and Food Chemistry* **54**(6): 2405-2411.

Hurlen, M., M. Abdelnoor, P. Smith, J. Erikssen and H. Arnesen (2002). "Warfarin, aspirin, or both after myocardial infarction." *New England Journal of Medicine* **347**(13): 969-974.

Ismail, H., E. Dilshad, M. T. Waheed and B. Mirza (2017). "Transformation of lettuce with rol ABC genes: extracts show enhanced antioxidant, analgesic, anti-inflammatory, antidepressant, and anticoagulant activities in rats." *Applied biochemistry and biotechnology* **181**(3): 1179-1198.

Ismail, H., E. Dilshad, M. T. Waheed, M. Sajid, W. K. Kayani and B. Mirza (2016). "Transformation of *Lactuca sativa* L. with rol C gene results in increased antioxidant potential and enhanced analgesic, anti-inflammatory and antidepressant activities in vivo." 3 Biotech **6**(2): 215.

Ismail, H. and B. Mirza (2015). "Evaluation of analgesic, anti-inflammatory, anti-depressant and anti-coagulant properties of *Lactuca sativa* (CV. Grand Rapids) plant tissues and cell suspension in rats." BMC Complementary and Alternative Medicine **15**(1): 199-203.

Jain, S. M. and P. M. Priyadarshan (2009). Breeding Plantation Tree Crops: Tropical Species.

John, T., R. D. Muller, A. Oberholzer, H. Zreiqat, B. Kohl, W. Ertel, A. Hostmann, S. K. Tschoeke and G. Schulze-Tanzil (2007). "Interleukin-10 modulates pro-apoptotic effects of TNF-alpha in human articular chondrocytes in vitro." Cytokine **40**(3): 226-234.

Kang, S.-R., J.-H. Choi, D.-W. Kim, S.-E. Park, K. Sapkota, S. Kim and S.-J. Kim (2016). "A bifunctional protease from green alga *Ulva pertusa* with anticoagulant properties: partial purification and characterization." *Journal of Applied Phycology* **28**(1): 599-607.

Khalid, S., N. Khalid, R. S. Khan, H. Ahmed and A. Ahmad (2017). "A review on chemistry and pharmacology of Ajwa date fruit and pit." *Trends in Food Science & Technology* **63**: 60-69.

Khan, F., F. Ahmed, P. N. Pushparaj, A. Abuzenadah, T. Kumosani, E. Barbour, M. AlQahtani and K. Gauthaman (2016). "Ajwa date (Phoenix dactylifera L.) extract inhibits human breast adenocarcinoma (MCF7) cells in vitro by inducing apoptosis and cell cycle arrest." *Plos one* **11**(7): 1-17.

Khan, H., S. Amin and S. Patel (2018). "Targeting BDNF modulation by plant glycosides as a novel therapeutic strategy in the treatment of depression." *Life Sciences* **196**: 18-27.

Khan, H., M. Saeed, A. U. Gilani, M. A. Khan, I. Khan and N. Ashraf (2011). "Antinociceptive activity of aerial parts of Polygonatum verticillatum: attenuation of both peripheral and central pain mediators." Phytother Res **25**(7): 1024-1030.

Khoo, L. T., F. Abas, J. O. Abdullah, M. Tohit, E. Rahayu and M. Hamid (2014). "Anticoagulant Activity of Polyphenolic-Polysaccharides Isolated from Melastoma malabathricum L." Evidence-Based Complementary and Alternative Medicine **2014**.

Kokras, N., D. Baltas, F. Theocharis and C. Dalla (2017). "Kinoscope: an open-Source computer program for behavioral pharmacologists." *Frontiers in Behavioral Neuroscience* **11**: 1-7.

Kripalani, R. and P. Kumar (2004). "Northeast monsoon rainfall variability over south peninsular India vis ‐ à ‐ vis the Indian Ocean dipole mode." International journal of climatology **24**(10): 1267-1282.

Lee, S., W. Lee, J.-S. Bae and E. Ma (2016). "Synthesis and in Vitro and in Vivo Anticoagulant and Antiplatelet Activities of Amidino-and Non-Amidinobenzamides." *Molecules* **21**(5): 676-685.

Lv, W. H., L. Zhang, S. J. Wu, S. Z. Chen, X. B. Zhu and J. C. Pan (2013). "Analgesic effect of ferulic acid on CCI mice： behavior and neurobiological analysis." China Journal of Chinese Materia Medica.

Malairajan, P., G. Geetha, S. Narasimhan and K. Jessi Kala Veni (2006). "Analgesic activity of some Indian medicinal plants." J Ethnopharmacol **106**(3): 425-428.

Mallick, N., R. A. Khan, A. Riaz and S. Afroz (2016). "Anticoagulant, Antiplatelet and Antianemic Effects of Citrus paradisi (Grape Fruit) Juice in Rabbits." *Pharmacology & Pharmacy* **7**(10): 397-400.

Mansouri, A., G. Embarek, E. Kokkalou and P. Kefalas (2005). "Phenolic profile and antioxidant activity of the Algerian ripe date palm fruit (*Phoenix dactylifera*)." *Food chemistry* **89**(3): 411-420.

Mathers, C. D. and D. Loncar (2006). "Projections of global mortality and burden of disease from 2002 to 2030." *PLOS Medicine* **3**(11): e442.

Mirza, M. B., F. Q. Syed, F. Khan, A. I. Elkady, A. M. Al-Attar and K. R. Hakeem (2019). Ajwa Dates: A Highly Nutritive Fruit with the Impending Therapeutic Application. Plant and Human Health, Volume 3, Springer: 209-230.

Mondal, H., S. Saha, K. Awang, H. Hossain, A. Ablat, M. K. Islam, I. A. Jahan, S. K. Sadhu, M. G. Hossain and J. A. Shilpi (2014). "Central-stimulating and analgesic activity of the ethanolic extract of *Alternanthera sessilis* in mice." BMC complementary and alternative medicine **14**(1): 398.

Mourão, P. A. and M. S. Pereira (1999). "Searching for alternatives to heparin: sulfated fucans from marine invertebrates." *Trends in Cardiovascular Medicine* **9**(8): 225-232.

Murata, T., F. Ushikubi, T. Matsuoka, M. Hirata, A. Yamasaki, Y. Sugimoto, A. Ichikawa, Y. Aze, T. Tanaka, N. Yoshida, A. Ueno, S. Oh-ishi and S. Narumiya (1997). "Altered pain perception and inflammatory response in mice lacking prostacyclin receptor." Nature **388**(6643): 678-682.

Mushtaq, Z., S. Kausar, N. Kousar and S. Chiragh (2017). "EFFECT OF AJWA DATE SEED ON LIPID PROFILE OF DIET INDUCED HYPERLIPIDEMIC RABBITS." Khyber Medical University Journal **9**(3).

Ojewole, J. A. (2005). "Antinociceptive, anti-inflammatory and antidiabetic effects of *Bryophyllum pinnatum* (*Crassulaceae*) leaf aqueous extract." Journal of Ethnopharmacology **99**(1): 13-19.

Patil, K. R., U. B. Mahajan, B. S. Unger, S. N. Goyal, S. Belemkar, S. J. Surana, S. Ojha and C. R. Patil (2019). "Animal Models of Inflammation for Screening of Anti-inflammatory Drugs: Implications for the Discovery and Development of Phytopharmaceuticals." Int J Mol Sci **20**(18).

Pietta, P.-G. (2000). "Flavonoids as antioxidants." Journal of natural products **63**(7): 1035-1042.

Poleszak, E., A. Szopa, E. Wyska, W. Kukuła-Koch, A. Serefko, S. Wośko, K. Bogatko, A. Wróbel and P. Wlaź (2016). "Caffeine augments the antidepressant-like activity of mianserin and agomelatine in forced swim and tail suspension tests in mice." Pharmacological Reports **68**(1): 56-61.

Rahmani, A. H., S. M. Aly, H. Ali, A. Y. Babiker and S. Srikar (2014). "Therapeutic effects of date fruits *(Phoenix dactylifera)* in the prevention of diseases via modulation of anti-inflammatory, anti-oxidant and anti-tumour activity." *International journal of clinical and experimental medicine* **7**(3): 483.

Ramsewak, R. S., M. G. Nair, S. Murugesan, W. J. Mattson and J. Zasada (2001). "Insecticidal fatty acids and triglycerides from *Dirca palustris*." Journal of agricultural and food chemistry **49**(12): 5852-5856.

Ramsewak, R. S., M. G. Nair, S. Murugesan, W. J. Mattson and J. Zasada (2001). "Insecticidal fatty acids and triglycerides from Dirca palustris." J Agric Food Chem **49**(12): 5852-5856.

Safayhi, H. and E. R. Sailer (1997). "Anti-inflammatory actions of pentacyclic triterpenes." Planta Med **63**(6): 487-493.

Saikia, D., R. Thakur and A. K. Mukherjee (2011). "An acidic phospholipase A2 (RVVA-PLA2-I) purified from *Daboia russelli* venom exerts its anticoagulant activity by enzymatic hydrolysis of plasma phospholipids and by non-enzymatic inhibition of factor Xa in a phospholipids/Ca2+ independent manner." *Toxicon* **57**(6): 841-850.

Sajid, M., M. R. Khan, S. A. Shah, M. Majid, H. Ismail, S. Maryam, R. Batool and T. Younis (2017). "Investigations on anti-inflammatory and analgesic activities of Alnus nitida Spach (Endl). stem bark in Sprague Dawley rats." *Journal of ethnopharmacology* **198**: 407-416.

Saryono, E. R., A. Proverawati and D. Hisni (2017). "Effect of antioxidant status and oxidative stress products in pre-menopausal women after treatment with date seed powder (*Phoenix dactylifera L.*): A study on women in Indonesia." Pakistan Journal of Nutrition **16**(6): 477-481.

Sheikh, B. Y., S. M. Zihad, N. Sifat, S. J. Uddin, J. A. Shilpi, O. A. Hamdi, H. Hossain, R. Rouf and I. A. Jahan (2016). "Comparative study of neuropharmacological, analgesic properties and phenolic profile of Ajwah, Safawy and Sukkari cultivars of date palm (Phoenix dactylifera)." Orient Pharm Exp Med **16**(3): 175-183.

Sheikh, B. Y., S. N. K. Zihad, N. Sifat, S. J. Uddin, J. A. Shilpi, O. A. Hamdi, H. Hossain, R. Rouf and I. A. Jahan (2016). "Comparative study of neuropharmacological, analgesic properties and phenolic profile of Ajwah, Safawy and Sukkari cultivars of date palm (*Phoenix dactylifera*)." Oriental pharmacy and experimental medicine **16**(3): 175-183.

Slattery, D. A. and J. F. Cryan (2012). "Using the rat forced swim test to assess antidepressant-like activity in rodents." *Nature Protocols* **7**(6): 1009-1014.

Stewart, A. M., Gaikwad, S., Kyzar, E., & Kalueff, A. V. (2012). Understanding spatio-temporal strategies of adult zebrafish exploration in the open field test, *Brain Research.*

Stone, W. M., B. H. Tonnessen and S. R. Money (2007). "The new anticoagulants." *Perspectives in Vascular Surgery and Endovascular Therapy* **19**(3): 332-335.

Tanelli[1], F., T. G. Belardo, C. H. F. Peiro, R. M. da Gama[1], F. L. A. Fonseca and J. Armando-Junior[1] (2017). "Preliminary Studies on Anticoagulation and Antiplatelet Activities of Plinia cauliflora (Mart.) Kausel Leaves Extract." *Journal of Applied Pharmaceutical Science* **7**(04): 115-118.

Tapas, A. R., D. Sakarkar and R. Kakde (2008). "Flavonoids as nutraceuticals: a review." *Tropical Journal of Pharmaceutical Research* **7**(3): 1089-1099.

Vayalil, P. K. (2002). "Antioxidant and antimutagenic properties of aqueous extract of date fruit (Phoenix dactylifera L. Arecaceae)." *Journal of Agricultural and Food Chemistry* **50**(3): 610-617.

Walter, U., M. Eigenthaler, J. Geiger and M. Reinhard (1993). Role of cyclic nucleotide-dependent protein kinases and their common substrate VASP in the regulation of human platelets. Mechanisms of Platelet Activation and Control, Springer: 237-249.

Wang, L., Z. Li, Q. Mei and D. Zhao (2000). "MN9202 protection of tail artery in carrageenan induced thrombos is rats." J Fourth Mil Med Univ **21**(2): 214-216.

Yan, F., J. Yan, W. Sun, L. Yao, J. Wang, Y. Qi and H. Xu (2009). "Thrombolytic effect of subtilisin QK on carrageenan induced thrombosis model in mice." *Journal of Thrombosis and Thrombolysis* **28**(4): 444-448.

Zhang, C.-R., S. A. Aldosari, P. S. Vidyasagar, K. M. Nair and M. G. Nair (2013). "Antioxidant and anti-inflammatory assays confirm bioactive compounds in Ajwa date fruit." Journal of Agricultural and Food Chemistry **61**(24): 5834-5840.

Zulfiker, A., M. M. Rahman, M. K. Hossain, K. Hamid, M. Mazumder and M. S. Rana (2010). "*In vivo* analgesic activity of ethanolic extracts of two medicinal plants *Scoparia dulcis L.* and *Ficus racemos* Linn." *Biology and Medicine* **2**(2): 42-48.

Zulfiker, H. M., M. M. Rahman, M. K. Hossain, K. Hamid, M. E. H. Mazumder and M. S. Rana (2010). "In vivo analgesic activity of ethanolic extracts of two medicinal plants-Scoparia dulcisL.andFicus racemosaLinn." Biology and Medicine **2**(2): 42-48.

Table 1. Antidepressant activity of rat induced *Phoenix dactylifera*

S. No.	Samples	Tail suspension test		Forced swimming test		Open field test				Hole board test					
		Struggling Time (s)	**Immobility Time (s)**	**Climbing Time (s)**	**Immobility Time (s)**	**Mobility Rate (%)**	**Explored Areas**	**Immobility Time**	**Number of Frozen Events**	**Mobility Rate (%)**	**Latency Time**	**No. of Head-dips**	**Explored Areas**	**Immobility Time**	**Number of Frozen Events**
1	PDF (300 mg/Kg)	69.32±2.2^{ns}	68.90±1.8***	76.98±3.8***	31.57±3.9***	75.95±1.8***	9.14±1.3^{ns}	73.50±4.5***	11.67±1.8***	45.39±4.1***	7.14±1.3***	7.00±1.4*	13.57±2.6***	11.29±3.9***	17.00±1.4***
2	PDF (200 mg/Kg)	65.31±2.1^{ns}	73.29±1.2**	47.63±2.1*	48.25±3.2*	65.24±1.1***	8.98±4.2^{ns}	82.25±2.5**	15.69±1.2***	36.28±3.1***	14.19±2.1***	5.25±2.3*	9.25±1.2***	18.67±3.2***	20.28±1.9***
3	PDF (100 mg/Kg)	61.11±3.3^{ns}	88.90±1.8***	34.98±1.4***	57.57±1.2***	57.91±1.2***	6.24±4.8^{ns}	95.22±1.5***	20.26±1.4***	22.96±5.5***	21.22±1.0***	4.91±3.8*	5.94±3.2**	27.24±3.3***	27.23±1.1***
4	PDS (300 mg/Kg)	89.70±3.3***	26.90±2.1***	88.20±1.7***	18.61±1.8***	81.76±1.4***	13.29±1.1**	54.71±4.3***	19.50±1.9***	50.09±4.9***	5.86±1.3***	9.29±2.6***	13.14±1.4***	9.86±1.9***	8.28±1.8**
5	PDS (200	78.75±4.	46.19±1.	62.42±3.	35.23±3.	72.62±3.	12.56±2.	63.26±1.	20.69±1.	39.66±4.	10.26±2.	7.92±1.2	9.19±3.9	16.57±1.	15.24±1.

	mg/ Kg)	1*	1***	5*	3*	2**	3**	3***	2***	1***	9**	***	*	4***	2**
6	PDS (100 mg/ Kg)	69.62±4.3*	67.21±1.5***	41.11±3.1*	56.28±4.1*	65.36±4.3**	10.11±2.1**	78.77±2.1***	24.29±1.3***	28.64±3.5***	15.99±3.4*	6.11±3.1*	5.91±2.1*	26.77±1.1***	23.61±1.1**
7	PDM (300 mg/ Kg)	87.31±4.8***	36.11±1.7***	78.39±1.7***	20.07±4.5***	73.38±2.5***	11.71±1.8^{ns}	87.71±3.4***	23.17±2.2^{ns}	50.57±4.7***	4.71±1.5***	6.29±1.8^{ns}	10.86±3.7**	13.14±3.8***	6.71±4.5^{ns}
8	PDM (200 mg/ Kg)	72.42±3.4*	44.15±1.2***	36.56±4.7*	36.41±2.4**	62.13±1.2***	10.17±4.4^{ns}	98.64±4.2*	26.12±2.4^{ns}	35.68±4.2***	9.78±2.1**	4.22±3.9^{ns}	7.32±1.4**	20.91±2.4***	13.79±3.5^{ns}
9	PDM (100 mg/ Kg)	70.41±3.2*	68.33±1.1***	29.12±4.1*	44.12±4.6*	58.23±4.5*	9.99±4.3^{ns}	107.23±3.8**	29.71±2.9^{ns}	24.97±3.4***	16.61±3.4**	2.24±4.4^{ns}	6.61±2.4**	29.26±4.2*	23.74±4.2^{ns}
10	PC (10 mg/ Kg)	93.71±4.4***	24.80±2.5***	88.33±3.6***	14.69±3.3***	90.52±1.6***	13.00±1.3**	28.43±4.7***	19.33±2.6***	60.91±3.7***	5.14±2***	10.29±1.7***	15.42±2.6***	9.43±1.7***	7.71±3.4**
11	PC (5 mg/ Kg)	85.51±2.4***	29.91±23.2**	72.32±1.2***	21.17±1.5***	85.29±1.2***	12.98±1.1**	32.68±3.4***	20.38±2.2***	56.98±4.0**	7.52±1.2***	9.36±1.2***	12.95±1.2***	11.32±1.1***	9.22±2.7**
12	PC (2.5 mg/	73.27±3.	35.23±3.	68.35±2.	29.71±2.	79.22±1.	12.52±1.	39.58±2.	21.22±2.	52.45±2.	9.31±2.1	8.78±2.1	10.11±2.	13.91±1.	10.23±3.

	Kg)	1***	6**	6**	1***	4***	0**	9***	1***	4***	***	**	0***	3***	1**
13	NC (10 ml/Kg)	67.63±3.1	95.10±4.6	26.60±2.9	60.74±2.3	55.95±0.2	10.14±1.3	132.98±0.6	25.00±1.8	14.96±3.8	24.17±1.7	4.14 ±1.3	5.00 ±2.2	31.57±3.7	3.71 ±1.5

Phoenix dactylifera Fruit (PDF), *Phoenix dactylifera* Seed (PDS), *Phoenix dactylifera* Fruit and Seed Mixture (PDM), Positive Control (PC; Fluoxetine-HCl), Negative Control (NC; Saline with 10% DMSO). Significant with * $p < 0.05$, ** $p < 0.01$, *** $p < 0.001$ and Non-significant (ns), values are expressed in Mean ± SD.

Table 2. Effect of analgesic activity of rat by writhes and hot plate assay.

S. No	Samples	Number of Writhes	Latency Time on Hot Plate Assay			
			0 h	1 h	2 h	3 h
1	PDF (300 mg/Kg)	$13.71 \pm 1.71^{***}$	7.43 ± 3.36^{ns}	$10.81 \pm 1.85^{***}$	$13.04 \pm 2.20^{***}$	$11.53 \pm 1.88^{***}$
2	PDF (200 mg/Kg)	$20.21 \pm 2.32^{**}$	7.21 ± 1.62^{ns}	$8.28 \pm 3.94^{*}$	$10.21 \pm 3.02^{**}$	$9.45 \pm 1.45^{***}$
3	PDF (100 mg/Kg)	$26.31 \pm 2.68^{**}$	7.63 ± 2.21^{ns}	$7.36 \pm 2.28^{**}$	$8.41 \pm 3.25^{**}$	$8.15 \pm 1.12^{***}$
4	PDS (300 mg/Kg)	$20.71 \pm 3.72^{***}$	6.67 ± 0.96^{ns}	$9.15 \pm 1.34^{***}$	$11.47 \pm 1.63^{***}$	$10.41 \pm 1.47^{***}$
5	PDS (200	$28.36 \pm 2.21^{***}$	6.10 ± 1.09^{ns}	$7.50 \pm 2.11^{**}$	$8.99 \pm 2.42^{**}$	$8.41 \pm 1.72^{***}$

	mg/Kg)					
6	PDS (100 mg/Kg)	37.72±4.24**	6.31±2.16^{ns}	6.85±2.21**	7.91±2.72*	6.69±1.40***
7	PDM (300 mg/Kg)	25.71±2.75***	7.49±0.79^{ns}	9.34±0.91***	12.01±1.35***	10.63±1.15***
8	PDM (200 mg/Kg)	33.63±1.52***	7.11±1.29^{ns}	7.94±1.01***	9.13±2.69**	8.31±1.61***
9	PDM (100 mg/Kg)	42.77±4.47**	7.61±0.37^{ns}	7.17±2.97**	8.45±2.53**	7.77±1.22***
10	PC (10 mg/Kg)	10.57±1.72***	6.43±0.69^{ns}	10.37±1.03***	11.80±1.32***	12.27±1.40***
11	PC (5 mg/Kg)	12.26±1.32***	6.39±1.92^{ns}	9.37±1.03***	10.67±2.11**	11.76±1.81***

12	PC (2.5 mg/Kg)	16.63±1.10***	6.64±2.11ns	7.37±1.03***	9.36±2.31**	10.18±1.68***
13	NC (10 ml/Kg)	74.57±2.64	6.67±0.84	7.08±0.84	7.16±0.74	7.58±0.68

Phoenix dactylifera Fruit (PDF), *Phoenix dactylifera* Seed (PDS), *Phoenix dactylifera* Fruit and Seed Mixture (PDM), Positive Control (PC; Morphine), Negative Control (NC; Saline with 10% DMSO). Significant with * $p < 0.05$, ** $p < 0.01$, *** $p < 0.001$ and Non-significant (ns), values are expressed in Mean ± SD.

Table 3. Determination of anti-inflammatory activity by paw Edema and histamine assay.

S. No.	Samples	Edema Volume (ml) in Paw Edema Assay				Edema Volume (ml) in Histamine Assay			
		0 h	1 h	2 h	3 h	0 h	1 h	2 h	3 h
1	PDF (300 mg/Kg)	$0.81\pm0.31^{*}$	0.59 ± 0.22^{ns}	0.36 ± 0.12^{ns}	$0.12\pm0.03^{***}$	$0.88\pm0.03^{**}$	$0.61\pm0.05^{***}$	$0.37\pm0.05^{***}$	$0.29\pm0.05^{***}$
2	PDF (200 mg/Kg)	$0.85\pm0.22^{*}$	0.64 ± 0.20^{ns}	0.40 ± 0.18^{ns}	$0.24\pm0.13^{**}$	$0.93\pm0.11^{*}$	$0.69\pm0.11^{*}$	$0.42\pm0.04^{***}$	$0.34\pm0.05^{***}$
3	PDF (100 mg/Kg)	$0.86\pm0.34^{*}$	0.71 ± 0.28^{ns}	0.49 ± 0.13^{ns}	$0.36\pm0.05^{***}$	$0.99\pm0.13^{*}$	$0.78\pm0.04^{***}$	$0.51\pm0.06^{***}$	$0.29\pm0.05^{***}$
4	PDS (300 mg/Kg)	0.49 ± 0.11^{ns}	0.31 ± 0.07^{ns}	$0.22\pm0.05^{*}$	$0.10\pm0.02^{***}$	0.90 ± 0.03^{ns}	$0.77\pm0.04^{**}$	$0.41\pm0.04^{***}$	$0.33\pm0.03^{***}$
5	PDS (200 mg/Kg)	0.54 ± 0.18^{ns}	0.49 ± 0.05^{ns}	$0.30\pm0.09^{*}$	$0.22\pm0.11^{*}$	0.92 ± 0.18^{ns}	$0.85\pm0.09^{*}$	$0.49\pm0.09^{**}$	$0.39\pm0.05^{***}$
6	PDS (100 mg/Kg)	0.55 ± 0.15^{ns}	0.58 ± 0.04^{ns}	$0.38\pm0.08^{*}$	$0.31\pm0.09^{**}$	0.98 ± 0.21^{ns}	$0.96\pm0.05^{**}$	$0.58\pm0.06^{***}$	$0.45\pm0.15^{*}$
7	PDM (300 mg/Kg)	0.73 ± 0.06^{ns}	0.33 ± 0.02^{ns}	$0.25\pm0.02^{*}$	$0.13\pm0.01^{***}$	0.91 ± 0.03^{ns}	0.59 ± 0.05^{ns}	0.33 ± 0.04^{ns}	$0.22\pm0.04^{***}$

8	PDM (200 mg/Kg)	0.79±0.16^{ns}	0.40±0.04^{ns}	0.31±0.09*	0.28±0.02***	0.95±0.06^{ns}	0.69±0.09^{ns}	0.36±0.05^{ns}	0.28±0.12*
9	PDM (100 mg/Kg)	0.78±0.04^{ns}	0.57±0.06^{ns}	0.43±0.05*	0.37±0.13*	0.94±0.02^{ns}	0.77±0.07^{ns}	0.39±0.03^{ns}	0.35±0.15*
10	PC (10 mg/Kg)	0.61±0.29^{n}	0.39±0.19^{ns}	0.29±0.15**	0.08±0.04***	0.93±0.04^{ns}	0.53±0.03***	0.22±0.03***	0.13±0.03***
11	PC (5 mg/Kg)	0.63±0.26^{ns}	0.45±0.18^{ns}	0.38±0.10**	0.15±0.06***	0.95±0.08^{ns}	0.59±0.12*	0.28±0.06***	0.19±0.04***
12	PC (2.5 mg/Kg)	0.69±0.21^{ns}	0.52±0.12^{ns}	0.45±0.25*	0.26±0.01***	0.94±0.05^{ns}	0.68±0.09**	0.31±0.07***	0.25±0.07***
13	NC (10 ml/Kg)	0.58±0.15	0.51±0.12	0.47±0.13	0.49±0.13	0.99±0.01	1.12±0.03	0.87±0.04	0.92±0.02

Phoenix dactylifera Fruit (PDF), *Phoenix dactylifera* Seed (PDS), *Phoenix dactylifera* Fruit and Seed Mixture (PDM), Positive Control (PC; Diclofenac Potassium for paw edema and Chlorpheniramine maleate for histamine assay), Negative Control (NC; Saline with 10% DMSO). Significant with * $p < 0.05$, ** $p < 0.01$, *** $p < 0.001$ and Non-significant (ns), values are expressed in Mean ± SD.

Table 4. Thrombus size in carrageenan induced thrombosis assay in rat by *Phoenix Dactylifera.*

S #	Sample	Thrombus size (mm)			
		0 h	1 h	2 h	3 h
1	PDF (300 mg/Kg)	15.57±2.79***	11.91±1.21***	4.80±1.31***	0.26±0.24*
2	PDF (200 mg/Kg)	16.42±3.57*	12.25±1.02***	5.53±1.72***	1.63±0.14*
3	PDF (100 mg/Kg)	16.24±2.48***	13.16±1.25***	6.63±1.46***	2.02±0.44*
4	PDS (300 mg/Kg)	16.43±3.51***	12.73±1.06***	5.51±1.66***	0.70±0.51ns
5	PDS (200 mg/Kg)	17.01±4.59*	13.45±2.10*	6.23±1.52***	0.85±0.59ns
6	PDS (100 mg/Kg)	16.54±4.84*	14.62±2.21*	7.13±2.64*	0.98±0.45ns
7	PDM (300 mg/Kg)	15.20±3.80***	13.13±1.31***	3.51±0.38*	0.52±0.44ns
8	PDM (200 mg/Kg)	16.56±4.51*	14.49±1.92**	4.93±1.01*	0.67±0.49ns

9	PDM (100 mg/Kg)	16.15±3.95**	15.192±1.95**	5.86±0.82*	0.79±0.35ns
10	PC (10 mg/Kg)	14.40±2.93***	10.26±1.43***	1.97±0.08***	0.31±0.09**
11	PC (5 mg/Kg)	15.26±1.84***	11.52±1.21***	2.52±0.81**	0.49±0.85*
12	PC (2.5 mg/Kg)	15.11±2.42***	11.99±1.13***	2.99±0.53**	0.64±0.72*
13	NC (10 ml/Kg)	20.23±4.28	20.23±1.41	9.26±3.24	2.00±1.61

Phoenix dactylifera Fruit (PDF), *Phoenix dactylifera* Seed (PDS), *Phoenix dactylifera* Fruit and Seed Mixture (PDM), Positive Control (PC; Aspirin), Negative Control (NC; Saline with 10% DMSO). Significant with * $p < 0.05$, ** $p < 0.01$, *** $p < 0.001$ and Non-significant (ns), values are expressed in Mean ± SD

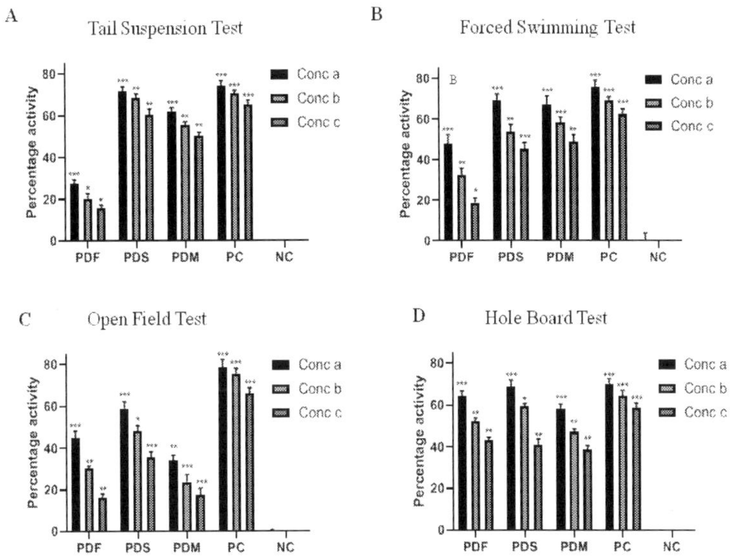

Figure 1 Antidepressant activity of *Phoenix Dactylifera.* (A) Tail Suspension Test (B) Forced Swimming Test (C) Open Field Test (D) Hole Board Test. *Phoenix Dactylifera* Fruit (PDF), *Phoenix Dactylifera* Seed (PDS), *Phoenix Dactylifera* Fruit and Seed Mixture (PDM), Positive Control (PC; Fluoxetine-HCl) and Negative Control (NC; 10 mL/kg saline in 10% DMSO). Statistically significant values expressed as * $p < 0.05$, ** $p < 0.01$, *** $p < 0.001$ and Non-significant (ns) in Mean ± SD. Where Conc a, b and c represents the three different concentration being 300, 200 and 100 mg/kg for plant extracts and 10, 5 and 2.5 mg/kg for PC.

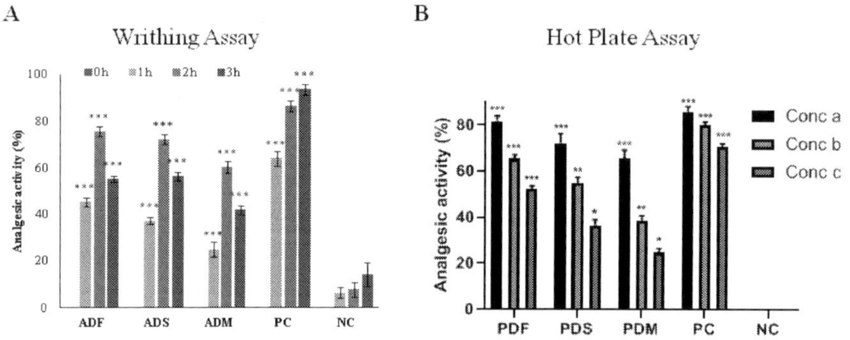

Figure 2 Analgesic activity of *Phoenix Dactylifera.* (A) Writhing Assay (B) Hot Plate Assay at the highest concentration (300mg/kg) is shown. *Phoenix Dactylifera* Fruit (PDF), *Phoenix Dactylifera* Seed (PDS), *Phoenix Dactylifera* Fruit and Seed Mixture (PDM), Positive Control (PC; Morphine) and Negative Control (NC; 10 mL/kg saline in 10% DMSO). Statistically significant values expressed as * $p < 0.05$, ** $p < 0.01$, *** $p < 0.001$ and Non-significant (ns) in Mean ± SD. Where Conc a, b and c represents the three different concentration being 300, 200 and 100 mg/kg for plant extracts and 10, 5 and 2.5 mg/kg for PC.

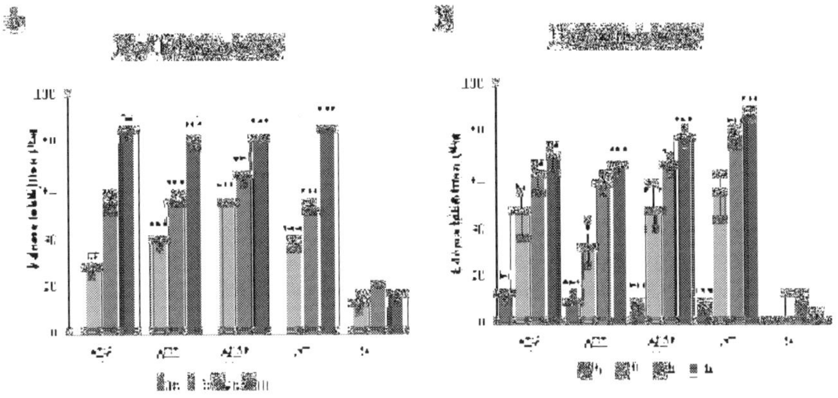

Figure 3 Anti-inflammatory activity of *Phoenix Dactylifera*. Percentage of Edema Inhibition (A) Paw Edema (B) Histamine Assay at the highest concentration (300mg/kg) is shown. *Phoenix Dactylifera* Fruit (PDF), *Phoenix Dactylifera* Seed (PDS), *Phoenix Dactylifera* Fruit and Seed Mixture (PDM), Positive Control (PC; Diclofenac Potassium for paw edema and Chlorpheniramine maleate for histamine assay) and Negative Control (NC; 10 mL/kg saline in 10% DMSO). Statistically significant values expressed as * p < 0.05, ** p < 0.01, *** p < 0.001 and Non-significant (ns) in Mean ± SD.

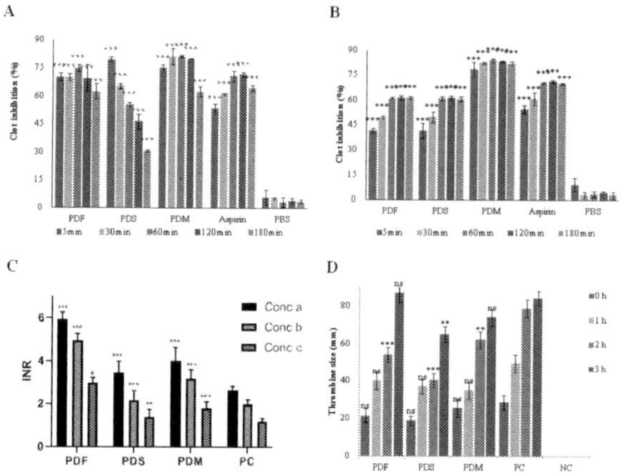

Figure 4 Anticoagulant activity of *Phoenix Dactylifera* extracts. (A) Percentage of clot inhibition in sheep blood assay (B) percentage of clot inhibition in antiplatelet assay (C) International Normalized Ratio (INR) in rat capillary assay (D) percentage of inhibition in carrageenan induced thrombosis assay at the highest concentration (300mg/kg) is shown. *Phoenix Dactylifera* Fruit (PDF), *Phoenix Dactylifera* Seed (PDS), *Phoenix Dactylifera* Fruit and Seed Mixture (PDM), Positive Control (PC; aspirin) and Negative Control (NC; 10 mL/kg saline in 10% DMSO). Statistically significant values expressed as * $p < 0.05$, ** $p < 0.01$, *** $p < 0.001$ and Non-significant (ns) in Mean ± SD. Where Conc a, b and c represents the three different concentration being 300, 200 and 100 mg/kg for plant extracts and 10, 5 and 2.5 mg/kg for PC.

Made in United States
Orlando, FL
04 November 2026

90200309R00026